Ligian Cristiano Gomes
Meri Lourdes Bezzi

The Spatial Dynamics and Materialisation of German Culture in Feliz/RS

Ligian Cristiano Gomes
Meri Lourdes Bezzi

The Spatial Dynamics and Materialisation of German Culture in Feliz/RS

Cultural plots

ScienciaScripts

Imprint
Any brand names and product names mentioned in this book are subject to trademark, brand or patent protection and are trademarks or registered trademarks of their respective holders. The use of brand names, product names, common names, trade names, product descriptions etc. even without a particular marking in this work is in no way to be construed to mean that such names may be regarded as unrestricted in respect of trademark and brand protection legislation and could thus be used by anyone.

Cover image: www.ingimage.com

This book is a translation from the original published under ISBN 978-3-330-77354-7.

Publisher:
Sciencia Scripts
is a trademark of
Dodo Books Indian Ocean Ltd. and OmniScriptum S.R.L publishing group

120 High Road, East Finchley, London, N2 9ED, United Kingdom
Str. Armeneasca 28/1, office 1, Chisinau MD-2012, Republic of Moldova, Europe
Printed at: see last page
ISBN: 978-620-7-99326-0

SUMMARY

ACKNOWLEDGEMENTS

Firstly, I thank God for life and for making everything happen at the right time.

- to the Federal University of Santa Maria for the opportunity to do my degree at a public HEI and the possibility of carrying out this research;

- To *Profª Drª Meri Lourdes Bezzi, who believed in my potential, for her patience in guiding me and contributing to my academic and personal growth, my eternal thanks;*

- *to the professors who made up the defence board for this graduation work, namely Professor Lauro Cesar Figueiredo and Professor Elizandra Voigt;*

- *To my colleagues at the Centre for Regional and Agrarian Studies (NERA/UFSM), especially Thales Silveira Souto, Paloma Tavares Saccol, Ricardo Stedile Neto, Andriele Prunzel Koglin, Elizandra Voigt and Helena Brum Neto, my eternal affection and thanks;*

- *the Municipality of Feliz for the information that was fundamental in carrying out this work;*

- *in particular, my mother Cleusa, my father Antônio, my sisters Lidiana and Liana, my stepmother Jussara, for their support throughout this stage and for teaching me to face the difficulties and challenges that life brings;*

- *to Mateus Pessetti for supporting me from the first day we met, even though we were far apart, he made my days shine so brightly;*

- *To my eternal friends and fellow undergraduates, especially Leticia Maffi Augusti, Jaqueline da Rosa Barreto, Micheli Gomes Stefano, who were there every step of the way, thank you very much;*

Finally, I would like to thank everyone who has been present in my life in some way and who has been essential to my personal growth.

CHAPTER 1

INTRODUCTION

[...] the place where one lives must be known and recognised by those who live there, because knowing the space, knowing how to move around in it, how to work and produce in it means being able to reproduce oneself as well, as a subject. This reality can be the city (or municipality), which is par excellence the shared territory, the place of life, where reproduction takes place in a given time and space, of the world that is global, of the universal. (CALLAI, 2004, p. 03).

Cultural studies have become the object of study for various sciences. However, in Geography, this theme has been studied since its inception and aims to explain human relations and their manifestations in space. Consequently, it establishes a diversity of themes to be investigated in the research carried out in this field of knowledge.

Geography's central concern is the society/nature interface in its various aspects. From the moment that culture became part of its studies, Cultural Geography was born. Since culture is a fundamental concept for Cultural Geography to describe the relationship in which man systematises with the environment. Culture, mediated by codes, is represented and materialised in space, giving rise to typical forms that can be recognised by other social groups.

According to Cuche (2002), the word culture comes from Latin and means the care of fields and livestock, appearing at the end of the 13th century to designate a plot of cultivated land.

In this sense, culture has become a key concept for geographical science, explaining the various relationships that man establishes with nature and the society in which he lives. It can be said, then, that culture consists of a set of actions by a social group, through a symbolic system, which is responsible for its identification.

They bear witness to the history of places and represent the cultural system that guides socio-spatial arrangements (VOIGT, 2013).

It is argued that culture also consists of a set of beliefs and values that guide the actions of different social groups. Cosgrove (1998) reinforces the importance of culture as an organiser of space, responsible for identifying that group.

The study of Cultural Geography is one of the ways of interpreting and analysing spatial organisation and/or reorganisation, as well as explaining the nature-society relationship. According to Wagner; Mikesell (2003, p. 29) "culture attributes meaning to everything from deliberately articulated vocal sounds to beings, objects and places".

In this sense, Cultural Geography is suitable for explaining the relationship that man establishes with his environment and its influence on the materiality of space. Understanding culture has become essential to understanding the symbolism inherent in each social group, since differentiation is measured by it (BRUM NETO, 2007).

Through the differentiation of people and places, through the characteristics employed in spatial use and occupation, there is an increase in the aspects that will identify each social group. In this sense, culture characterises the peculiarities of each space.

Brazil has a great cultural diversity due to its heterogeneous ethnic background. In this context, this study will focus on the municipality of Feliz/RS, which is characterised by the significant presence of German descendants, both in rural and urban areas (FIGURE 1).

It should be noted that the population of the municipality under study maintains the cultural traditions of their ancestors, preserving and developing German material cultural codes, which are materialised in the landscape, especially represented by architecture, typical festivals, gastronomy and religiosity. The immaterial code to be analysed is the dialect of the German language, i.e. orality, which is commonly used among its inhabitants as a way of maintaining their cultural unity. In this sense, these elucidated codes will be responsible for the cultural structuring of this research.

In this sense, Feliz was chosen as a research target due to its socio-historical-cultural potential, with the aim of contributing to the dissemination of its tourism potential, as well as the knowledge and traditions of its people.

Figure 1 - Location of the Municipality of Feliz/RS

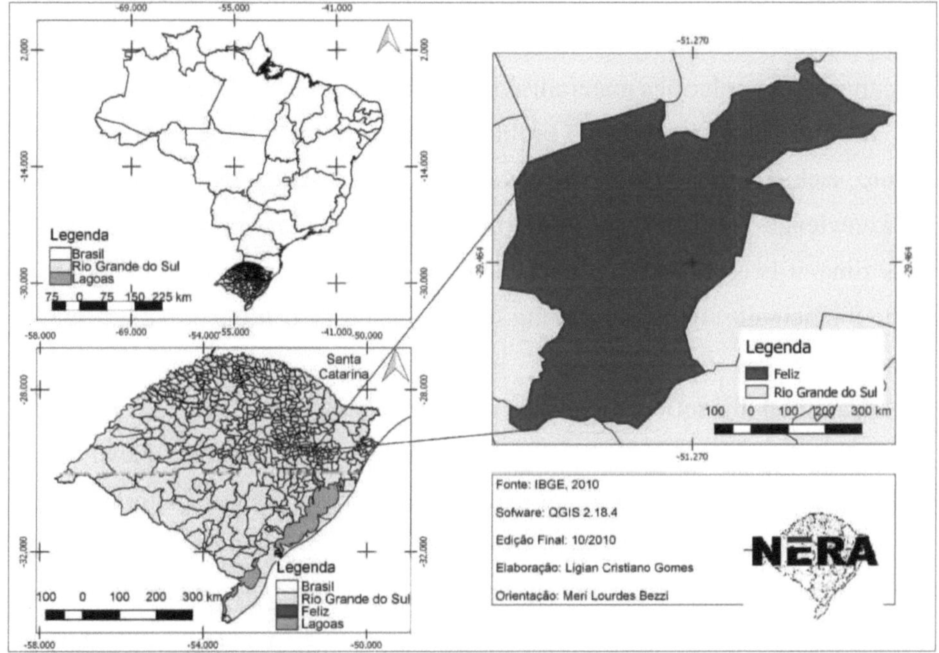

Org: GOMES, L. C. (2017).

It should be noted that the territorial unit under study has a population made up of different cultures, totalling 12,359 inhabitants, according to the Brazilian Institute of Geography and Statistics (IBGE, 2017).

The municipality under study is located in the Vale do Caí region, which is characterised by providing an education that stands out from other regions.

municipalities and for its excellent quality of life, which is recognised nationally.

This territorial unit stands out culturally for its location in the state of Rio Grande do Sul, situated between the two main economic centres of the state: the metropolitan region (80 km from Porto Alegre) and the mountainous region of the state (45 km from Caxias do Sul). The city has easy access and links to the main state and federal motorways, such as RS 122, RS 240 and BR 116

(PREFEITURA MUNICIPAL DE FELIZ, 2017).

The relevance of the research is justified by the need to identify Germanic cultural materialisations and, consequently, the contribution of this ethnic group to local spatial formation, with the aim of preserving memory and disseminating the cultural particularities materialised in the municipality's landscape.

In this sense, the research problem aims to reflect on the local space and culture, seeking to understand the development of its spatial dynamics as well as the knowledge and activities of German immigrants and descendants. At the same time, it is concerned with analysing how cultural codes have materialised in the municipality of Feliz, seeking to understand their transformations and/or permanence.

Based on this reflection, in order to demonstrate that the object of study of geographical science is the nature-society interface, it is essential to analyse the different forms of appropriation of nature by the German ethnic-cultural group.

The intrinsic objectives of the research were: (a) to analyse the concepts of space and culture, as well as the cultural contributions of ethnic Germans to the spatial formation of Feliz/RS; (b) to identify cultural materialisation through material codes, such as architecture, gastronomy and typical festivals, and the immaterial ones, especially the orality of ethnic Germans present in this spatial area; and (c) to point out the main codes materialised in the landscape of the municipality under analysis.

The methodology established for the development of the work was organised using a flowchart of the research structure's procedures. Through its stages, it sought to meet the proposed objectives. In this way, the first stage consisted of a theoretical and methodological in-depth study, by means of a historical and cultural review, looking at specific bibliographies to identify the main concepts guiding the theme in question.

The following authors were used as bibliographical sources for the first topic: ALMEIDA (2008), ASSMANN (2009), BRUM NETO (2004, 2006; 2007), BEZZI (1996), CLAVAL (1999), COSGROVE (1996, 2007), CORRÊA,

ROSENDAHL (2001, 2007), CUCHE (2002), HAESBAERT (1999, 2007), REGO (1995), MCDOWELL (1996), ZANATTA (2011).

In the second stage, data was collected from secondary sources, such as: census data available on the website of the Brazilian Institute of Geography and Statistics (IBGE), Feliz City Hall, historical documents available in the Historical Archive of the territorial unit under study, as well as from Cultural Research organisations, Museums, the Department of Culture of the municipality of Feliz, since these bodies hold fundamental information for carrying out the research.

At the same time, another fundamental phase of the research was carried out in the field, which corresponded to the third stage of the research. This became vital in order to highlight the cultural aspects present in the landscape, i.e. those that contain the "cultural imprint", expressed through the cultural codes of ethnic Germans. It should also be noted that during this stage it was possible to identify the main codes materialised in the landscape of the municipality under analysis, at which point photographs were also taken to demonstrate the cultural codes still present in the area under study.

The last stage of the research corresponded to the analysis and interpretation of the results, making it possible to perceive the materialisation of Germanic cultural codes, as well as the analysis and understanding of German cultural insertion in the local context, considering the cultural perspective, its contribution and perpetuation to the spatial dynamics of the territorial unit investigated.

After the defence of the Bachelor's Degree in Geography, articles will be written and published in journals dealing with this subject, with the aim of disseminating the experience gained from this scientific research.

CHAPTER 2

REVIEWING THEORETICAL MATRICES

> Culture is the sum of behaviours, knowledge, techniques, knowledge and values accumulated by individuals during their lives and, on a different scale, by all the groups they belong to. Culture is an inheritance passed down from one generation to the next. It has its roots in the distant past, which plunges into the territory where their dead are buried and where their gods manifest themselves (CLAVAL, 1999, p. 63).

This chapter presents the theoretical matrix, emphasising the guiding concepts for the research. In this context, the perspectives that emphasise the understanding of Culture, Cultural Geography, Cultural Codes and Cultural Identity are highlighted.

2.1 THE CONCEPT OF CULTURE

With regard to the concept of culture in Cultural Geography, it should be emphasised that it remains impregnated in the inherent approaches to this holistic theme. However, regardless of the fact that culture is a field shared jointly with the human sciences, it has generous connections, since each social group perceives and transmits its respective space in a unique way.

The word culture therefore externalises numerous techniques.
Social groups are usually associated with the study and knowledge inherent exclusively to individuals, where they interpret each other. In addition to this common fact, culture is everything that establishes meaning in the world that surrounds certain individuals or social groups.

Cosgrove (1996, p.53) emphasises that,

> The concept of culture, in terms of beliefs and values shared by a human group, only emerges with the recognition of alternative belief and value systems, and under the motivation of a "disinterested" study. In modern societies, this has led to the increasing recognition of the relativity of cultural truths. Alongside this relativism has come an interest in subjecting modern cultures to critical analysis and recognising that they are made up of a plurality of voices that construct meaning for the world in different ways.

The evolution of this concept allowed it to become more categorical as its use and presence in debates progressed in numerous sciences, especially Geography, in proportion to their theoretical and methodological bases.

It is emphasised that each social group has its own culture, with its own distinctive marks and characteristics, which is called Cultural Identity. Consequently, even if other cultures are understood, the cultural identity of these groups will remain the same, as it influences their upbringing, way of acting and way of thinking.

From this perspective, Zanatta (2011, p. 7) emphasises that

> The opening up of new horizons for analysing the geographical dimension of culture was found in the revaluation of fundamental characteristics of humanism. Thus, man was put back at the centre of cultural geographers' concerns, as the producer and product of his own world.

It is important to emphasise that culture has become a key concept for Cultural Geography, fostering the relationship that man establishes with the environment, as opposed to its perpetuation with space. In this way, culture is fundamental to analysing and understanding the countless singularities of each social group.

Based on these reflections, we can say that the ethnic-demographic flows that occupied Rio Grande do Sul endowed the territory with unique cultural characteristics, with significant changes to the traditions that had existed up to that point, inserting their codes and customs (BRUM NETO, 2004).

The contribution of different cultures, including the European one, was not restricted to the economic sphere, but took on greater proportions. In this sense, it can be considered that the process of ethnic-cultural occupation was responsible for the development of an original civilisation, that is, the gaucho civilisation, because there was a

process of assimilation of the codes that already existed in the region by these new inhabitants, who contributed in the same way, incorporating their habits and customs into gaucho soil, represented by the specific cultural codes of each ethnic group. Through these aspects, regionalism became the most expressive "brand" and also a form of appropriation of space, giving it an identity (BRUM NETO, 2004).

From these concepts, we can emphasise that man is a producer and product of his culture, materialising relevant creations in space, using symbolic processes that determine his ethnic-cultural identity, establishing his cultural codes. According to Brum Neto (2007, p. 38) cultural codes are,

> [...] in the symbolism responsible for the visibility of culture and also for its transmission. They are imprinted in the different landscapes, through the style of the houses, typical clothing, art, gastronomy, music, religiosity and festivities. In addition to these, there are other codes which, although not visible, are also responsible for materialising culture in space, such as cultural contributions, especially values, ideologies and conventions. In this process of cultural codification, communication, both oral and written, stands out as one of the essential codes for transmitting and projecting culture in time and space.

In literature reviews on the subject, according to Bezzi (1996, p. 236) "[...] it is through the study of customs and habits that we can interpret a particular social group and perceive the regional disparities that guide spatial analyses of development". In this way, culture emerges as a way of elucidating the organisation of space, where the countless social groups that materialise on it end up creating new and distinct spatial realities.

In this line of reasoning, Rego (1995, p. 58) where the author establishes that development is closely related to the sociocultural context in which the person is inserted and is processed dynamically (and dialectically) through ruptures and imbalances that provoke continuous reorganisations on the part of the individual.

In this way, it can be understood that culture distinguishes social groups, which in turn end up transfiguring the cultural codes pertinent to their ethnicity in the landscape, emphasising the material and immaterial principles of culture. It should be emphasised, therefore, that the studies that cover this sphere of geographical science introduce us to the multiple forms that social groups find themselves in.

According to Corrêa; Rosendahl (2003, p. 18) it should be noted that

We can understand, based on the numerous surveys elucidated, that Culture aims to broadly foster objects that address the organisation/reorganisation of space, through cultural codes, which are materialised in the landscape, thus addressing the sense of understanding between human groups that are perpetuated in this environment. However, such approaches are emphasised by Claval (1999), inferring it with the development of Cultural Geography.

In this way, the studies that Geography still carries out on different parts of the globe show that the forms of occupation have been unequal. Culture and space or culture-space occupy a central position in the scientific investigations carried out, among other sciences, by Geography (CAETANO, 2012).

2.2 CONCEPTS OF CULTURAL GEOGRAPHY

Cultural Geography focuses its studies on the organisation of space, where it is the inherent product of culture materialised and immaterialised by cultural codes.

In this way, Mikesell (2000, p. 88) conceptualises that "[...] the growth of cultural geography is cumulative rather than additive. It is basically achieved through intercultural comparison, historical verification and the expansion of research from smaller to larger areas."

In this sense, Sauer (2007, p. 23) emphasises that

The development of cultural geography necessarily proceeds from the reconstruction of the successive cultures of an area, starting with the original culture and continuing up to the present. The most rigorous work carried out to date refers less to current cultural areas than to previous ones, since these are the foundation of the present and their combination provides the only basis for a dynamic vision of the cultural area. If cultural geography, engendered by geomorphology, has one fixed attribute, it is precisely the evolutionary orientation of the subject.

The cultural question, based on studies that touch on Cultural Geography, has a wide variety of themes addressed in Brazilian scientific research, since Brazil has a great cultural diversity that is pertinent to its heterogeneous ethnic background. In this way, the current challenges of cultural studies are characterised by the relationship between culture and multinationalisation, a worldwide phenomenon that promotes greater inter-relationship between cultures, reducing the original cultural characteristics of social groups.

Almeida (2008, p. 47) emphasises that

> It can be seen that geographers were concerned with understanding the landscape from the point of view of man's actions, his relationship with the elements of nature and/or the transformed nature, and made a cultural reading through practices, ignoring men's attitudes and beliefs.

In this sense, geography in its multiple studies that improve the concept of culture is elucidated by numerous authors, thus broadly emphasising that the concept of culture is relatively old and originates in France, where it materialises the conception of man's existence. In this sense, Brum Neto (2006, p. 4) points out that "[...] along the investigative path traced by Geography, culture has been widely debated, initially in Germany (with Ratzel), France (La Blache) and the United States (Sauer) and subsequently spreading to various countries".

From these concepts, we can emphasise that Cultural Geography in Germany and the United States investigated the materialist perception of culture. In this way, Brum Neto (2006, p. - - - emphasises that "The development of Cultural Geography in Germany and the USA was perceived to have followed the same path, i.e. it only considered the material part of culture, i.e. what is visible in space. However, it neglected cultural knowledge and values".

In this line of reasoning, Corrêa (1999, p. 51) understands that the renewal of the theme broadens the studies, as follows

> The resurgence of cultural geography, after a period of relative loss of prestige between 1940 and 1970, meant, both in Europe and the United States, a thematic renewal and, more than that, a renewal of the approach. [...] The resurgence of cultural geography takes place in a post-positive context of spatial organisation and its dynamics. The cultural dimension becomes necessary for understanding the world.

Based on these reflections, it can be emphasised that since these events, countless authors have begun to highlight and address cultural issues in their interpretations, emphasising the material and immaterial aspects of culture.

Thus, we emphasise that the themes researched by Cultural Geography after 1970 can be inferred from the approaches to the Geography of Religion, which ended up being consolidated through the numerous studies carried out by the Centre for Studies and Research on Space and Culture (NEPEC), which began in 1993 at UERJ in Rio de Janeiro.

Corrêa (1999, p. 53) points out that "The themes of religion, environmental perception, spatial identity and the interpretation of texts (literature, music, painting and cinema) are among the other themes that emerged or were taken up again".

According to Claval (2001, p.78) it is important to note that

> The cultural geography of the early 20th century was often criticised for paying too much attention to the past, for its ignorance of contemporary societies and for its inability to deal with current social situations. The new cultural approach is much more critical. It often focuses on present situations, the struggles they characterise and the problems of social justice: in a way, it continues to bear the mark of the critical radicalism that inspired it at its inception. But the new cultural approach is above all attentive to the meanings that people give to the cosmos, nature, the environment and society. It also explores the religious dimensions of spatial experience and the role of ideologies in modern secular societies. The humanist and phenomenological roots of the new orientation are therefore always active. However, its essential significance lies elsewhere. The new cultural approach makes it possible to restore coherence to the discipline by moving away from a relativistic perspective. A certain predictability in social behaviour is introduced by the experience of communication.

Among the many authors who share the idea that Cultural Geography has undergone a transformation, it is worth highlighting that from the 1970s onwards, the introduction of new perspectives and thus, with the deepening of studies, the internal debates of Geography would be rethought and in turn deepened in theoretical-methodological issues.

In this context, Claval (1999, p. 63) emphasises that

> The renewal of cultural geography is due to two factors. The discipline is confronted with new ways of affirming the diversity of groups, which it cannot ignore. The work

<blockquote>
of epistemological reflection undertaken by the social sciences and geography since the early 1960s has reached a decisive point. The inconsistencies of the positivist principles hitherto accepted have been realised.
</blockquote>

However, Cultural Geography has undergone a process of renewal as a result of numerous criticisms, thus seeking out new approaches.

have enriched this line of research. From these surveys, it can be concluded that the appreciation of culture becomes concrete, emphasising material and immaterial aspects, thus highlighting the corresponding symbolism of each culture.

According to McDowell (1996, p.159) it should be emphasised that

<blockquote>
Cultural geography is currently one of the most exciting areas of geographical work. Ranging from analyses of everyday objects and the representation of nature in art and film to studies of the meaning of landscapes and the social construction of place-based identities, it covers numerous issues. Its focus includes the investigation of material culture, social customs and symbolic meanings, approached from a range of theoretical perspectives.
</blockquote>

In this way, it can be said that the influence of historical and dialectical materialism is shown through the conception of culture as both a representation and a social condition, where the recognition of the historical dimension is perpetuated in the relationship between man and nature.

2.3 CULTURAL CODES

A culture materialised in space through cultural codes is characterised as a set of symbols. They prioritise the perpetuation of that culture. The codes, in turn, transfer the cultural characteristics of a given social group through the generations.

In this sense, these cultural characteristics are embodied in the landscape, through the architectural style of the houses, gastronomy, dances, festivities and the typical German dialect, cultural materialities/imaterialities of the municipality under study. It should be noted that in addition to these, there are other codes that, although not visible, are also responsible for the materialisation of culture in space, such as cultural contributions, especially values, ideologies and conventions (BRUM NETO, 2007).

The transfer of customs enables the materialised perpetuation of the original cultural codes of a social group, specifying the evolution of the symbolic sets characteristic of a given culture, remodelling them to their new reality. This emphasises the significant relationship between the culture manifested in space by the innumerable particular forms of cultural groups. "Culture results from the ability of beings to communicate with each other through symbols".

In this context, Cosgrove (2007, p. 103) emphasises

> Human beings experience and transform the natural world into a human world through their direct engagement as thinking beings with their sensory and material reality. The production and reproduction of material life is necessarily a collective art, mediated in consciousness and sustained through communication codes. The latter is symbolic production. These codes include not only language in its formal sense, but also gesture, dress, personal and social behaviour, music, painting, dance, ritual, ceremony and constructions. Even this list does not exhaust the series of symbolic productions through which we maintain our lived world, because all human activity is both material and symbolic, production and communication. This symbolic appropriation of the world produces distinct lifestyles (genres de vie) and distinct landscapes, which are historically and geographically specific. The task of cultural geography is to learn and understand this dimension of human interaction with nature and its role in the ordering of space.

In this sense, the material and immaterial aspects of culture pertinent to a social group become responsible for identifying typical characteristics, attitudes and paths established by these groups, establishing the perpetuation of their values, customs and ideologies.

Values can be considered to be abstract beliefs and norms of behaviour, generally the domain of religion and metaphysics. Ideologies are rational ideologies that give meaning to history and guarantee the social order of peoples (CLAVAL, 1999).

Cultural codes are a set of representations in which the behaviour of each individual highlights the value and relevant actions of each group, thus establishing their customs and knowledge.

Still on the subject of religion, it can be emphasised that it can be considered the most striking code within ethnic groups, expressing the value that belief materialises for the people. According to Claval (1999, p.115) "sharing the same religious or metaphysical beliefs and taking part in the same rites that bring believers together constitute very solid social foundations".

In this way, religion can be seen as a point of cultural unity, based on the common beliefs of ethnic cultural groups. They have the same beliefs and share the same rituals from birth to death as the individuals who make up the social group. However, rituals vary according to culture, as beliefs differ. Generally, religiosity serves as a guide for collective actions, setting a standard to be followed (BRUM NETO, 2007).

According to Mikesell (2000, p. 90) "In most parts of the world, geographical cultural research implicitly includes knowledge of the isolation or mixture of languages and dialects". This highlights the importance of also considering slang, popular sayings, nicknames and language expressed by the body as a characteristic part of a culture, showing that this cultural code is therefore manifested spontaneously and also through school instruction via literacy.

Mikesell (2000, p. 29) emphasises that

> [...] Exclamations, gestures, facial expressions etc. are also languages; otherwise, paintings, emblems and everything that is regularly recognised as "signifying something" are also languages. Finally, objects and behaviours of all kinds enter into the communication process. Culture attributes meaning to everything from deliberately articulated vocal sounds to beings, objects and places.

From these meanings, we can emphasise that the transmission of customs typical of an ethnic group portrays the perpetuation of the culture's original cultural codes, pertinently highlighting the evolution of the symbolic system characteristic of this culture, and therefore establishing models to adapt to the new reality.

It can therefore be said that cultural codes are established as an alternative for ethnic groups, enhancing their culture through the transfer of typical symbols, where they become identifiers in relation to other social groups. In this way, each culture has its own essential symbolic set for understanding; however, these characteristics will undergo changes over time, evolving as these groups move forward with reality.

2.4 CULTURAL IDENTITY

The term cultural identity is established as a typical symbol that ends up characterising a particular culture. In this way, cultural identity is perpetuated between the difference in symbolic sets, which represent the forms of each social group.

In this way, the term identity is first perpetuated in relation to the feeling of belonging to a social group, where the "sensation" of sharing with the environment is materialised. This establishes the contemplation of one's own existence.

From this perspective, Brum Neto; Bezzi (2008, p. 29) emphasise that

> [...] cultural identity is indispensable for the maintenance of a social group and also for understanding the production of local and regional space at their respective levels of development. Today, identity has become a factor in "valorising" space as a way of boosting the local/regional economy through différance. In a globalised world, culture has become the differentiator that results in socio-economic attraction and often provides a prospect for development.

We can understand from the authors that cultural identity values the space in which it is materialised, where personal experience is taken into account and promotes local/regional development. In this sense, the real principle predominates in the elements that guide all individuals related to the same culture.

According to Claval (1999, p. 98), we can infer that

> Identity is at once individual and collective, tastes and experiences vary in each person, but the internalisation that makes the values to be respected conscious during adolescence tends to impose the same shape on the image of oneself.

From these concepts, we can point out that every culturally inferred element in a collective environment provides its idealisation of the space, which forms a material code and that which does not form such a visualisation is established as an immaterial code of culture[1] . Identity is thus the most consolidated relationship between human beings, their societies and their spaces.

Brum Neto (2007) emphasises that cultural identity serves as a distinction between groups, based on difference. It is the result of the relationship between a social group and its spatial base, through the establishment of links. For Cuche (2002), there is no identity in itself, nor for itself, but always in relation to another, accompanying difference.

In this sense, we emphasise that although they are related, these concepts have different meanings, where Cuche's conception (2002, p. 176) refers to the fact that

> [...] culture can exist without awareness of identity, while identity strategies can manipulate and even modify a culture that will then have little in common with what it was before. Culture depends to a large extent on unconscious processes. Identity refers to a norm of attachment, necessarily conscious, based on symbolic oppositions.

It is important to emphasise that there is a link between culture and cultural identity, which allows a link to be established between their concepts. This can be established on the basis that culture is fostered by the essence/nature of a social group. He goes on to emphasise that cultural identity is consolidated in the feeling of belonging or not belonging to an ethnic cultural group in a given territory.

Still on this subject, it should be noted that the established cultural and social meaning presupposes that identity perpetuates a characterisation between what is similar and what is different or, at the same time, difficult to distinguish.

In this sense, Haesbaert (2007, p.3) emphasises that

> As any concept, "culture" and "identity" are defined through the contractive preposition with other concepts, which give them a meaning that is not exactly

---

[1] It should be noted that this study will focus on characterising the material attributes of a culture first, and then the immaterial ones.

Generally speaking, identity originates from the codes that identify the culture and are therefore decisive. Once the codes are established and the identity is built, it begins a process of consolidation over time, where its codes are constantly tested. They can thus remain if they are "solid" enough, or disappear if they prove fragile. They can also be replaced by others, or even added to with new elements and undergo reformulation and/or resignification (BRUM NETO, 2007).

According to Hall (1997, p. 42) it should be considered that

From these perceptions, we can emphasise that each ethnic group has helped to structure the national cultural identity, and that this help is underpinned by a specific cultural code. However, this contribution does not narrow down the nation culturally, where heterogeneity transcends geographical space.

One of the most important characteristics of territorial identity is that it draws on a historical dimension, from the social imaginary, so that the space that serves as a reference "condenses" the group's memory (HAESBAERT, 1999).

These reflections point us towards the pluralism that makes culture "rich", as it demonstrates the coexistence of diversity forming national unity. Immigrants retain many traits of their culture of origin, but also add new codes, the result of a natural evolutionary process. Communities emerge with diverse cultures, with plural identities, originating from an ethnic-cultural mosaic that cannot be attributed to a single ethnic origin (BRUM NETO, 2007).

Based on these surveys, it is worth highlighting another important aspect of identity, where Claval (1999, p.98) emphasises that

Identity is at once individual and collective. Attitudes, tastes and experience vary in each person, but the internalisation that makes one aware of the values to be respected during adolescence tends to impose the same shape on the image one makes of oneself. In certain cultures, the desire to fully realise oneself is valued. But by wanting to fulfil oneself so much, one risks forgetting the prescriptions that collective life demands.

In this line of reasoning, the understanding of each social group is made up of a subject, who is impregnated by a certain cultural heritage. In this context, their interaction with the other members of a community becomes homogenised, thus triggering the collective ethnic and cultural development that will perpetuate this group.

Thus, we understand that the ways in which a people identify themselves, through symbolic systems of representation, give them legitimacy in relation to others, making them appropriate an identity that allows them to become distinct and original based on a common cultural essence (BRUM NETO, 2007).

It should be noted that there will always be typical characteristics that denote a social group, resulting in a recognisable identity. Cultural homogenisation, resulting from the globalisation process, is established as a mediator, since the standardisation of the way of life will find resistance in the elements that involve the cultural essence.

CHAPTER 3

CHARACTERISATION OF THE GERMAN IMMIGRATION PROCESS IN BRAZIL: RIO GRANDE DO SUL AS A FOCUS

> Without knowing for sure what awaited them here, the immigrants certainly remained focused on the fact that life in a country with so many attractions would be less arduous than in their mother country. For these people, having a roof over their heads to shelter their families and having land to plant and grow their food, with a peaceful lifestyle, was the most anticipated event (GABBI, 2014, p.14).

This chapter emphasises the process of German migration in Brazil, focusing on the resulting dynamics in the state of Rio Grande do Sul. It discusses the first group of Germans who came to Brazil, the first colony, as well as the structure of German-Brazilian immigration and its socio-spatial/cultural (re)organisation in Rio Grande do Sul.

3.1 INSERTION OF GERMAN CULTURE IN BRAZIL AND RIO GRANDE DO SUL

To approach German immigration in Brazil, a broad theoretical discussion is required. The peculiarities resulting from this process led to the socio-spatial organisations of the municipalities that received this social group. During the period that mediated this colonisation process, the current spatial organisation found in the state of Rio Grande do Sul took shape.

In this sense, it should be noted that in May 1824, Emperor Dom Pedro I was responsible for the first wave of German immigrants to Brazil. The start of the journey across the Atlantic Ocean was the beginning of a gruelling adventure. The immigrants landed in areas close to Rio de Janeiro, with the aim of creating agricultural colonies to supply the city. However, due to the influence of Empress Leopoldina, preference was given to colonising the southern region of the country (ASSMANN, 2009).

The beginning of the 19th century saw the first boats sail across the Atlantic Ocean. These were three-masted sailing ships equipped with cannons due to the fear of attacks by pirates. With regard to passenger conditions, the precariousness of the crossing stands out. From Hamburg to Rio de Janeiro, the immigrants had to endure a long journey, which could take more than a hundred days (ASSAMANN, 2009).

It should be noted that the first group of German immigrants to Brazil settled in the southern region of Bahia. Their first colony, however, was established in the far south of the country, in São Leopoldo, which is now part of the Porto Alegre metropolitan area. The main reason for their immigration was the search for a quality of life, a condition not found at the time in their country of origin (ASSMANN, 2009).

In order to understand the structure of the immigration of German-Brazilians, the formation of the first colonies and, subsequently, the understanding of the process of socio-spatial/cultural organisation in Brazil, an important figure must be highlighted. It should be noted that the forerunner in the process of colonising descendants of Germanic origin in the south of the country, especially Rio Grande do Sul, was Dona Leopoldina. The Empress's priority was the colonisation of this territory by immigrants of purely German origin. In this way, the Germans were motivated by the conquest of the land and the expectation of a prosperous life. In this sense, countless immigrants left the so-called Old World.

In this context, Brum Neto (2007, p. 124) points out that

> [...] the German-Brazilian population flow served the interests of both, since in Germany there was a large human contingent, ravaged by misery and driven from their lands by wars. In Brazil, large tracts of land remained unexplored and ready to be populated.

The vast majority of German immigrants who landed in Brazil were peasants dissatisfied with the loss of their land, former artisans, free labourers and entrepreneurs wishing to exercise their activities freely, political persecutors, people who had lost everything and were in difficulty, migrants who were "hired" through incentives to manage the colonies or who were hired by the Brazilian government to work at an intellectual level or to take part in battles (ASSMANN, 2009).

However, it was during the 20th century that most German immigrants arrived in Brazil. More than 70,000 Germans arrived in the country in the 1920s. Most of these would no longer go to the rural colonies, but to the urban centres. The city of São Paulo received most of this new wave of German immigration, with around 20,000 Germans living in the city in 1918. Others went to Curitiba, Porto Alegre and Rio de Janeiro. There were also groups of Germans in the 1940s who emigrated to Brazil because of the Second World War (ASSMANN, 2009).

The numerous interests of the empire were not only alienated from the colonisation process, but with the aim of sending soldiers to the Corps d'Estrangers in Rio de Janeiro, whose purpose was to militarily guarantee independence, which Lisbon denounced as a "simple rebellion". (ASSMAM, 2009).

However, the immigration of Germans was part of the general motivations behind European emigration, which were driven by political, economic, social and cultural transformations. The development of industrial capitalism and the subsequent end of feudal ties established an environment of population aversion in the face of an open frontier and a utopian America.

Following this line of thought, Pesavento (1992, p. 157) points out that

> For an immigrant country like Brazil, the phenomenon is linked to the fundamental moment in which the transition from slave labour relations to wage labour relations took place at national level. This is the internalisation of capitalism in Brazil, when historically, capital began to take ownership of production on a global level. Fundamentally, the foreign immigrant was intended to provide labour to replace the slave labour force on the coffee plantations.

German emigration, like all European emigration, is explained by the major socio-political and economic reorganisations that Europe underwent in the 19th century. The consolidation of the German national state was crucial to the growth of emigration. In addition, Brazil experienced exceptional conditions in the 19th century that favoured European immigration. In the second half of this century,

European immigrants arrived in São Paulo to provide labour for the coffee plantations and to supply peasants for the colonial settlements that were being created by the Brazilian government. One characteristic of this process is the ticket that immigrants received when they boarded the ships (IBGE, 2017).

In 1824, the state of Rio Grande do Sul established the process of immigration of Germanic peoples to its territory. This process of German immigration took place during the migratory movement of the 19th and 20th centuries. The genesis of immigration was based on a number of aspects, such as the frequent social problems that were occurring in Europe. Another aspect that can be highlighted is the issue of the "whitening of the Brazilian population", which was one of the factors behind the immigration process to Brazil (ASSMANN, 2009).

The origin and regional composition (TABLE 1) of the German immigrant groups depended to a large extent on the criteria and preferences of the emigration agents in Germany, while their destination in Brazil was left in the hands of the Brazilian recipients who distributed them, taking into account skills, geopolitical and economic interests. The diversity and cultural heterogeneity of the groups of Germans who arrived in Brazil in the 19th century was remarkable. They came mainly to settle in colonies in the south-east and south of the country, where the imperial government established the colonies of São Leopoldo (RS), São Pedro de Alcântara and Mafra (SC) and Rio Negro (PR). Still in the 19th century, German settlers were also taken to other regions of the country, such as Espírito Santo, Minas Gerais, Rio de Janeiro and Bahia (IBGE, 2017).

Table 1 - Origin of some immigrant groups to southern Brazil

Where some German groups came from in southern Brazil		
Location	Foundation	Origin
São LeopoldoRS	1824	Hunsruck, Saxony, Wurtteerg, Saxony-Coburg
Santa Cruz RS	1849	Renãma. Pomerâma. Silesia
Santo ÂngeloRS	1857	Rhineland. Saxony Pomerania
Nova PetröpolisRS	1859	Pomerania. Saxony, Bohemia
Teutônia RS	1868	Westfalia
Sfio Lourenço RS	1857	Pomerania. Renãma
BlumenaulSC	1850	Pomerania Holtein, Hannover. Braunschweing. Saxony
Search.SC	1860	Baden Oldenburg Rhineland Pomerania, Schleswig-Holstein Braunschweig
Joenville/SC	1851	Piússica, OWenburg, Schleswig Holstem, Hannover Switzerland
Curitiba PR	1878	Volga Theutes
Santa Isabel-[1] ES	1847	Hunsruck, Pomerania, Rhineland, Prussia, Saxony
São LeopoldinaES	1857	Pomerania Rhineland. Prussia Saxony

Source: IBGE, Brazil 500 years, 2017.
Org: GOMES, 2017.

From this perspective, it should be noted that there were other reasons for the colonisation process during this period. These included the need for soldiers for the army, due to the reality of the Province of Rio Grande do Sul, which due to its geographical location had its territory bordered by a border region and constant combat against the Platinos[2] .

Referring to these processes, Brum Neto (2007, p. 125) reports that in Rio Grande do Sul

> [...] there were areas predestined for the establishment of colonies, in the wastelands, the immigrant did not choose their location. It should be emphasised that the way colonisation was carried out in the province, with the initial predominance of Germans, with insertion in continuous portions of space, meant that there was a certain homogeneity from a cultural point of view.

There is also another major factor that can be cited to justify this immigration: the replacement of slave labour. During this period, slavery was in a progressive process of decline. Although the replacement was necessary, the process of free and salaried labour was not to the liking of the large coffee growers in south-eastern Brazil, who wanted to maintain the structure of slave exploitation with these new immigrant populations (IBGE, 2017).

In light of this, the imperial government decided to set up a department solely to take care of the services inherent in the colonisation process. At the time, German Major Jorge Antônio Von Shaeffer was appointed to look after this new service. It was up to him to start the process of hiring German settlers for this department. In this regard, Pellanda (1925, p. 3-4) points out that the first step was the immediate production and distribution of printed bulletins promising those wishing to emigrate the following advantages:

> 1° To pay the fares of Germans who wanted to come and colonise Brazil 2° To admit German settlers to the Empire as Brazilian citizens, whose status they would enjoy as soon as they arrived.
> 3° No impediment to the worship professed by the settlers, whatever it might be - this freedom, it was said, was guaranteed to them by the Constitution of the Empire.
> 4° To give each settler and each head of family a free and clear plot of land,

---

[2] A colonial period in which Argentina, Paraguay and Uruguay all took part in the same administration, where the countries had a history of conflict and co-operation with Brazil (ASSMANN, 2009).

measured and demarcated, with a surface area of 160,000 square braças (approximately 77 ha), partly in fields, partly in farmland, and partly in virgin forest.
5° Grant horses, oxen, cows, sheep, pigs, etc. as free property to each settler or in proportion to the size of their families.
6° To pay each settler daily, during the first year, the sum of one franc (160 réis) and in the second year, half (80 réis) per head, indistinctly.
7) For the first ten years, settlers will be exempt from paying duties, both on their income and on any other object, and during this time they will be exempt from any state service.
8° The settlers would receive everything free of charge and as free property, but they could not dispose of any of it for the first ten years - after which they could dispose of their property by paying tithes on the produce of their crops.
9° The settlers were also obliged to formally renounce their nationality of origin.

Against this backdrop, the advantages for immigrants willing to colonise the land were extraordinary, as settling in the lands of the New World had never been so tempting. Leaving their lives behind, selling their property, getting rid of personal belongings and taking what was really reusable began a long journey into the new unknown.

Despite all the benefits on offer, one question was asked: why did German immigrants want to leave their land? The answer is the same as that given to any other migration process carried out by humankind: they were looking for better living conditions.

The first German settlers arrived in Rio Grande do Sul on 25 July 1824. They settled on the banks of the Rio dos Sinos, where the municipality of São Leopoldo now stands. At first, the group consisted of just nine families, totalling 39 people. Between 1824 and 1830, around 5,350 immigrants arrived in the state (ASSMANN, 2009).

However, within six years the immigration process was interrupted, but from 1844 to 1850 more than 10,000 immigrants began to arrive in the state again and, consequently, until 1889 the same number of German immigrants was repeated in this new period of immigration to Rio Grande do Sul. By 1914, it was estimated that 17,000 Germans had arrived in the state. During this same period, there were already more than 50,000 Germans in Rio Grande do Sul, where around 140 colonies were set up throughout the state (ASSMANN, 2009).

The colonies began to expand across the state, most of which were located near or on the banks of rivers, as strategic points for receiving and transporting produce. However, in order to perpetuate all these processes, the immigrants had to adapt to the natural conditions of the land they had received. Thus, they perfected the art of wielding the axe and specialised in cutting and burning the native forest.

The state of Rio Grande do Sul underwent other colonisation attempts, these in places that were difficult to access, where one of the aims of this new colonisation was to send groups of people who were causing some kind of problem, making them undesirable in São Leopoldo. This new attempt took place in the Missões region, in the colony of São João das Missões. The other colonisation attempt took place on the coast, in the Torres region, with the aim of populating the forest zone between Rio Grande de Sul and Santa Catarina. However, both attempts failed (ASSMANN, 20090).

With regard to the countless influences that German immigrants introduced into our culture, we can mention the implementation of new agricultural techniques, the improvement of craft techniques, the crucial point being the manufacture of footwear, the metal industry, textiles and so on.

In this context, it is clear that numerous European immigrants were instrumental in promoting the organisation of new socio-economic, political and cultural structures in southern Brazil. The heterogeneity between immigrant groups can be explained by the fact that they did not come from the same region and thus did not share the same socio-cultural heritage, which in turn was very marked, to the point where the German immigrants themselves questioned themselves as foreigners (IBGE, 2017).

It should be noted that the Germans are very representative of the other ethnic groups that migrated to the country. They are second only to the Italians, who are responsible for the main ethnic/cultural formation in the southern region of Brazil.

Therefore, it should be noted that the cultural essence of immigrants played an important role in shaping Brazilian ethnicity/culture, and especially that of the state of Rio Grande do Sul, especially with regard to certain eating habits, typical theatre performances, church choirs, music bands and so on.

Another example of the inclusion of the cultural habits of these people is the festivals, particularly the Oktoberfest, which has German origins and was introduced to Brazil due to the need for immigrants to feel a sense of belonging. It should be emphasised that this festival emerged in Brazil as a way of demonstrating against the attitudes of the Estado Novo, which banned cultural activities that identified Germanness. Today, this event is a festival that symbolises German culture, incorporating, with adaptations and modifications, German food, music and language (IBGE, 2017).

With regard to the reproduction of the handicraft culture in Brazil, it should be emphasised that this came about through the habits that these immigrants had when they lived in Germany. This practice was developed in that country because the winter period was very harsh, making it impossible to work outside the home. Crafts were then produced inside their homes, valuing and stimulating the large family unit that came from the German immigrants.

Another point that stands out in this phase concerns the particular characteristics of the colonies, where there were predominantly groups of the same ethnicity, and in the workshops there was a concern to mix cultural elements from different origins.

In this sense, the perpetuation of the Brazil/Rio Grande do Sul colonisation process is noteworthy, and the typical characteristics of this ethnic group are present and materialised on the time scale that marked this process of German immigrant colonisation in the state. In this sense, the next chapter will address the pertinent issues of Germanic insertion in the territorial unit under investigation.

3.2 PECULIARITIES OF THE GERMAN ETHNIC GROUP AND THE INSERTION OF THIS GROUP IN THE MUNICIPALITY OF FELIZ

Studies and work on the preservation of cultural assets have been developed and focused on a recent period in Brazil. In this way, discussions about the inherent aspects that structured the immigration process in the national territory have established peculiar marks in the space, especially in terms of culture and the economy. The importance of valuing the past and collective memory configures colonisation as a remarkable process of cultural exploration/implantation.

Approaches to the cultural formation of certain regions are essential to understanding the current configuration of ethnic cultural formation. In this way, Cuche (2002, p. 28) emphasises that,

> Culture comes from the soul, from the genius of a people. The cultural nation precedes and calls forth the political nation. Culture appears as a set of artistic, intellectual and moral achievements that constitute a nation's heritage, considered to be definitively acquired and the foundation of its unity.

Considering the process of cultural formation of the southern region of Brazil, it is worth emphasising that the presence of different groups, such as the native groups of the region at first, such as the Indians, the European groups highlighting the Spanish, Portuguese and later Italians and Germans, were fundamental in promoting singularities as well as ethnic cultural peculiarities.

In relation to the formation and socio/political/economic and cultural development of Rio Grande do Sul, it can be seen that in different periods the immigration process had different values for the growth of the state. However, during the Farroupilha Revolution (1835-1845) there was an interruption in immigration to the state. However, after the conflict ended, the imperial government resumed the process of German immigration.

One of the first immigrants, Dr Hillebrand, was appointed director of the São Leopoldo colony. In order to enlarge the colony, Dr Hillebrand had the land between the Forromeco stream and the Caí river measured in 1846. These lands belonged to the imperial government, so they were measured by Dr Hillebrand and distributed to the settlers. The lands located on the left bank of the Caí River belonged to the region of Escadinhas, where the Moraes farm existed. The towns of Picada Cará, São Roque, Coqueiral and Bananal belonged to Tristão Monteiro. Just upstream, in the direction of Nova Petrópolis, there was the Pirajá farm, whose name still exists. These lands on the left bank of the River Caí were acquired in the 1860s by German settlers who bought them from these farmers (ASSMANN, 2009).

It can be inferred that the Bugres[3] , the Caingangues[4] and the Botocudos[5] inhabited caves and the forest in that region. In this respect, it is worth noting that, today, it is possible to find objects in caves that perpetuate the existence of bugres in that place.

The genesis of German immigration to the region began in 1846, with the pioneering immigrant Dr Hillebrand. During this phase, there were families of other ethnicities in the Gaucho campaign, so families of German origin were directed to the Vale dos Sinos, Caí and Taquari (ASSMANN, 2009).

In this process, these lands were not ceded to be occupied, but with the precept of serving as a social and political heritage in the future, according to the letter from engineer Afonso Mabilde to Dr Hillebrand found in the book Feliz ontem e hoje (Happy Yesterday and Today) (MABILDE, 1860, p.22).

> Your Lordship well knows what it costs for a family to support itself in a colony on mountainous land that does not exceed 16,000 square metres, and especially when it has married children who also farm on the same land. Therefore, the settlers of Picada Feliz are beginning to murmur, because they know that there are land titles on the right bank of the River Caí, and this is as far up as the colonies that have been populated until now. Your Lordship is not unaware that the Picada Feliz, for a long distance parallel to the river Caí, is no more than half a league long. What I can assure you is that a large part of the settlements on the Picada Feliz are established on land that was conceived many years ago and of which there are titles. Therefore, it is impossible to continue distributing land in Picada Feliz until the government makes a decision on the matter.

It should be noted that this region was covered in vast forest, making it impossible to occupy it immediately. The river Caí was a land divider. The banks

---

[3] The name given to indigenous people who were considered non-Christian by Europeans (ASSMANN, 2009).

[4] The Caingangues, Kainguangs, Kaingang, Kanhgág, Guainás, Camés or Xoclengues establish themselves as an indigenous people and the 5th most numerous of the five indigenous peoples in Brazil today (IBGE, 2017).

[5] Botocudos is the generic name given by the Portuguese colonisers to the different indigenous groups belonging to the macro-Jê trunk (non-Tupi group), from different linguistic affiliations and geographical regions, whose individuals wore lip and ear botoques (IBGE, 2017).

on the left side of the river were occupied by landowners who exploited the forest to produce charcoal. However, from 1860 onwards, the land on the left side of the river was bought by German settlers.

At this point, it should be noted that in 1853, Feliz had approximately 90 families. Most of them were from other colonies (ASSMANN, 2009). It should be noted that the colonisation process faced numerous difficulties, due to the various attacks that this group faced from other peoples.

In this sense, according to Wagner and Mikesell (2003, p. 28)

> [...] it no longer considers isolated individuals or any personal characteristics they may possess, but communities of people occupying a given, broad and generally continuous space, as well as numerous characteristics of belief and behaviour common to the members of such communities.

It should be noted that the whole process of colonisation and development of these peoples in this region was marked by numerous problems and difficulties. Initially, they were deceived by the false promises of those who promoted immigration. Another difficulty was the annual invasions of locusts on their crops, resulting in losses and damage. A cloud of these insects would settle over the crops and, in two or more days, wipe out the plants, leaving only the stems bare of leaves. The solution to this terrible problem was only found when the government distributed a highly toxic insecticide, locust powder (Hochrequenqueft), which was sprayed while the locusts were having their last meal. It is reported that the last locust invasion took place more than 60 years ago (ASSMANN. 2009).

It should be noted that the perpetuation of the colony was being established, but it needed a link to the fields of Vacaria and the colony of São Leopoldo. Thus, the construction of a road by engineer Mabilde was realised after thirty years. It is clear that throughout this period, the colony of Feliz faced difficulties in consolidating its trade and selling surplus products, as access to the town was long and complicated. As a result, they only produced products for their own consumption.

Based on these surveys, it can be emphasised that due to this process of lack of outlets and self-consumption of production, leisure activities were developed in the colony. Activities such as card games, bowling, bocce, target shooting, among others, gave rise to the habit of drinking during this process.

At another point, the construction of the road linking São Leopoldo to the north of the state (Campo dos Bugres and Vacaria) became the only access route to the town. This road linked Feliz to São Sebastião do Caí, via Escadinhas. At this time, Italian colonisation began. It is worth noting that Feliz soon became a route for the economy of the north-eastern region at that time. This led to the development of hotels, in particular, as there was a large influx of travellers and traders (ASSMANN, 2009).

In this sense, it is worth noting that during this period the colony had all its transport structured around animal-driven carts.[6] It's worth noting that during rainy periods, the cart drivers[7] faced difficulties, having to wait up to a month to be able to cross the rivers with their carts on improvised rafts until the water level receded. It should be emphasised that many products were lost due to this long wait.

As a result, it became necessary to build a bridge over the River Caí at Picada Feliz. In November 1898, the brothers João and Aquelo Corrêa Ferreira da Silva were promised the construction of a bridge at Picada Feliz.[8] The construction concession was also given to the Corrêa brothers for a period of 25 years, during which they or their direct heirs could charge tolls. After this period, the bridge would pass into public ownership in good condition (ASSMANN, 2009).

This explains the history of the iron bridge (FIGURE 3) in the Feliz

---

[6] It should be noted that horses, oxen and donkeys were kept in the Feliz colony. It should be noted that these animals were present throughout the colonial period, since the population and goods were transported by them in the mid-1898s (ASSMANN, 2009).

[7] A "Carreteiro" is a man who transports products, groceries and various items by ox cart (IBGE, 2017).

[8] The town of Picada Feliz has 500 inhabitants. The community has a church, village hall and three cemeteries. Picada Feliz also has two municipal schools. Most of the population are tobacco and dairy farmers and also cultivate maize, beans, potatoes, soya, onions and other crops (IBGE, 2017).

landscape. It was manufactured and brought from Belgium, where it was used as railway transport. Thus, in March 1900, fifteen months after its manufacture began, it was inaugurated. It should be emphasised that this was a milestone, resolving all the problems the town was facing with the railway crossing. As a result, on 16 June 2016, the Feliz iron bridge was listed as a State Historic Site.

Still on this subject, it should be noted that after 190 years of use, the Iron Bridge needed to be refurbished, which would increase its useful life by many years. The municipal administration (2009-2012) sought funds from the Ministry of National Integration to carry out a major renovation. In total, R$288,986.97 was invested, of which R$140,836.99 came from the Ministry and the other R$148,149.97 was released by the municipality. During the work, the asphalt concrete roadway and concrete pedestrian walkways were replaced with raised steel plates. The change reduced the weight of the bridge by 78 tonnes, giving it a longer service life. The structure was also overhauled and improved, with the guardrails reinforced and the headwalls re-paved, as well as the construction of a wall. Thus refurbished and completed, the municipal administration handed over the Iron Bridge to the community of Feliz on 9 December 2009 (ASSMANN, 2009).

Figure 3 - Representation of the start of construction and later completion of the Iron Bridge in Feliz
Source: Prefeitura de Feliz, 2017.

In this context, we can see the care and importance that Feliz attaches to the origins of its history. In this way, points characteristic of the initial colonisation process are preserved and the perpetuation of culture in the territorial unit under study remains alive among the inhabitants. In this way, the origin of the municipality's name is emphasised, where ASSMANN (1902, p. 26) points out that

> The name Feliz is attributed to a historical event where, in 1850, an entourage under the command of the German engineer Afonso Mabilde went to the city.
>
> tasked with opening a route through the pine forest and Campo dos Bugres (Caxias do Sul) to the cattle ranching fields of Vacaria. This group crossed the Rio das Antas in a canoe, using a boat as a link to the necessary supplies. However, a flood washed the canoe away and the group of men were forced to return south. After many days wandering through the bush suffering all sorts of hardships and dangers, they finally found a settler's house and greeted this encounter with the exclamation: O happy one! In memory of this fact, the new road was given the name "FELIZ".

It is argued that since the beginning of its colonisation, Feliz has been characterised by the principles of its people, who have been seen as providers of the value of work, health and education for its inhabitants. It's important to note that these aspects contributed to Feliz being ranked first among the Brazilian municipalities with the highest human development index (HDI) in 1998[9] , according to the report published by the United Nations (UN)[10] . (ASSMANN, 2009).

As a result, according to estimates by the Brazilian Institute of Geography and Statistics (IBGE) 2017, the population of Feliz is 13,273 inhabitants, some of whom are rural and others urban. In this sense, it should be noted that in terms of ethnic origin, the population is made up of 70% German, 15% Italian and 15% of other origins such as Polish, Portuguese, Swiss, Austrian, among others (IBGE, 2017).

When we look back at the past, we can see that culture promotes a differentiated study of space, based on the concepts that characterise a social group. Culture, mediated by cultural codes, is represented and materialised in space, giving rise to typical forms that establish an understanding of this territorial unit. In this way, the next chapter will highlight the cultural codes of the Germanic ethnic group materialised in the municipality of Feliz.

---

[9] It is summarised as a measure of long-term progress, established through three basic dimensions of human development: income, education and health (UNDP, 2017).

[10] It is an intergovernmental organisation made up of 192 sovereign states and founded after the Second World War[a] to maintain peace and security in the world, foster friendly relations between nations and promote social progress in order to promote international cooperation (UN, 2017).

CHAPTER 4

CULTURAL CODES: THE MATERIALISATION OF GERMAN CULTURE IN THE MUNICIPALITY OF FELIZ/RS

> [...] cultural geographers share the same goal of describing and understanding the relationships between collective human life and the natural world, the transformations produced by our existence in the world of nature and, above all, the meanings that culture attributes to its existence and its relationships with the natural world (COSGROVE, 2000, p.34).

This chapter will address the Germanic cultural issue present in the municipality of Feliz and the spatial organisation resulting from this process. It emphasises the cultural codes present in the landscape, both tangible and intangible.

4.1 THE MATERIALISATION OF GERMAN CULTURE IN THE MUNICIPALITY OF FELIZ

In this respect, when addressing the central objective of Geography's studies and analyses, it is worth highlighting the society/nature interface. From the moment that culture became part of the studies of this science, Cultural Geography originated, being understood/explained as a kind of subfield of Geography.

For geographers who work with cultural issues and seek to analyse the dynamics of society, it is essential to identify peculiar aspects arising from the process of socio-spatial development/increment/transformation. In this sense, in order to better understand the peculiarities and singularities of certain ethnic/cultural aspects, it is important to identify cultural codes. These should be understood in two ways: material and immaterial codes. Material cultural codes are visualised through the architectural style of houses, music, religion, festivals and clothing, and immaterial ones, such as norms, beliefs, values and ideologies, among others (CAETANO, 2012).

Cultural codes identify a social group. In this sense, the German ethnic group and the characteristics that are intrinsic to this social group are emphasised. Thus, culturally, the German-Brazilians are remembered as hard-working and enterprising people. Through the colonisation of immigrants from Germany in southern Brazil, especially Rio Grande do Sul, they took their characteristics with them, transforming their "new land" into their own space.

The municipality of Feliz links its cultural codes to the space, origin and memory of its people, perpetuating the pertinent peculiarities of the Germanic ethnic group in this territorial unit under study. In this way, Feliz materialises its cultural codes, knowledge and activities in space, highlighting its historical/ethnic/cultural relevance. There is a sense of appreciation for cultural identity in the municipality, represented with great significance by the inhabitants' sense of belonging.

This idea gains strength when Ruviaro (2011, p. 23) emphasises,

> [...] to strengthen the community and visitors' awareness of the importance of preserving architectural heritage, as an attraction of great importance for the development of tourism practices (through the signposting of some examples), making them recognise and preserve it by rescuing the events and facts of these places of memory [...].

Feliz's cultural codes are evident throughout its spatial area, allowing for a broad analysis of its corners, streets and avenues. This territorial unit presents a diversity of symbologies that materialise German culture. The cultural manifestations are alive and reproduced by its immigrants from Germany, thus establishing the process of settling in the territory, which have been transmitted, taught and perpetuated over the generations.

The Germanic cultural issue is significant in the spatial organisation of Feliz. During the process of investigation and development of the research, the immigrants' pride in maintaining cultural traditions and their cultural identity, passed down through the generations, can be seen. This shows that the ethnic roots are/will be present and perpetuated over time.

In this respect, it should be noted that the material cultural codes present in the territorial unit under study include typical German architecture. The façades of some of the buildings and houses are reminiscent of the Enxaimel style[11] , a style of construction in which wood is fitted together. Another characteristic is the construction of the roofs, which are robust and steeply sloping, preventing the accumulation of snow.

It should be noted that in Germany, Enxaimel construction developed where there was a great abundance of suitable and more resistant hardwoods, in other words, throughout the German plain and the middle plateau in the centre of Germany. The wood most commonly used was oak[11] [12] , which at this time was still found in large quantities in these regions. But oak is characterised by its slow growth, requiring a period of 200 years before it can be felled. For this reason, it began to die out and was replaced by spruce wood[13] and beech wood[14] . As Germanic culture has always had a very sedentary village population and contacts between different regions were difficult and infrequent, the spread of the Enxaimel was not linear (WEIMER, 2005).

It should be emphasised that these characteristics of typical German architecture are reminiscent of the classic building style of their ancestors. The reproduction and perpetuation of this material code of Germanic culture is kept alive in the municipality of Feliz. This can be seen in the buildings that reproduce and materialise the typical features of this architecture.

The town hall is built in the Enxaimel style, which harks back to the

---

[11] Experts and builders believe that the Enxaimel originated in the region of present-day Germany during the Middle Ages. It is considered to have originated when houses stopped using piles buried in the ground and started using stone or masonry foundations. This was even influenced by the Roman occupation of former Germania. From this moment on, houses began to last for generations and evolve technically, including various structural and cultural elements (COLÉGIO DE ARQUITETOS, 2017).

[12] This is the common name for species of trees in the genus Quercus of the Fagaceae family. The genus is native to the northern hemisphere and includes both deciduous and evergreen species that range from high latitudes to tropical Asia and America (FERREI RA, 1986).

[13] This is the popular name for the various species of the Abies genus. They are coniferous trees of the Pinaceae family, native to the temperate forests of Europe, Asia and North America (FERREIRA, 1986).

[14] It is the common name given to various species of trees, including the genus Fagus (Eurasian and North American beeches) (FERREIRA, 1986).

German colonisation of Feliz. The façades of other buildings located in the centre of the urban area also follow this architectural pattern. With these factors, the process of reproduction, as well as concern for the historical and cultural preservation of Germanic features, is perpetuated in the landscape (FIGURES 4 and 5).

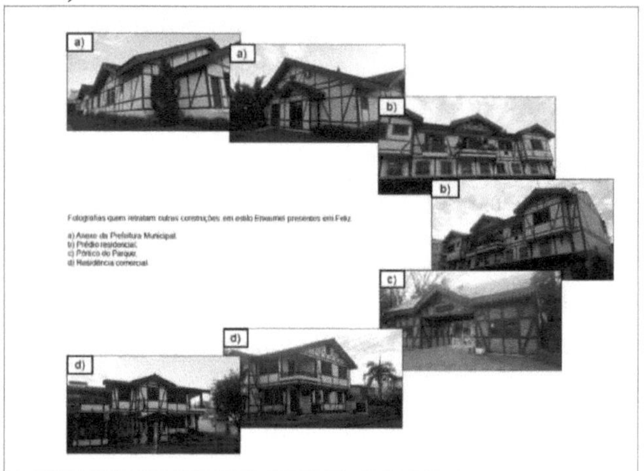

Figure 4 - Mosaic of photographs of typical Germanic architecture in the various buildings in Feliz/RS
Source: Fieldwork (2017).
GOMES, L. C. (2017).

5 - Mosaic of photographs from the Feliz/RS City Hall showing Enxaimel architecture
Source: Fieldwork (2017).
GOMES, L. C. (2017).

In this sense, the architectural representation typical of this social group is present and materialised. This can be seen in the mosaic of photos taken during the fieldwork carried out for this research. This provided a better analysis/understanding of the guiding factors that structure the historical-cultural processes responsible for the material codes of the Germanic ethnic group in Feliz.

Another material code of culture that can be elucidated are the typical German festivals. The National Beer Festival stands out in the town. This festival takes place annually in the municipality as a tribute to and a revival of the city's brewing tradition. The festival is organised by the Feliz Cultural and Sports Association - SOCEF (FIGURE 6).

In 1966, the director of the Polka Brewery in Feliz, Mr Victor Ruschel, travelled to Germany and had the opportunity to take part in the Ocktoberfest. Enthusiastic about what he had experienced, he decided to introduce a similar festival in the municipality, which he called the Feliz Chopp Festival. (MUNICIPAL PREFEITURA, 2017).

The Festival emphasises the essence of German culture, with the presence of typical bands, Germanic dinners and performances by the Feliz German Folk Dance Group.

In this context, in order for the first Chopp Festival to have characteristics similar to those of the Ocktoberfest, it was necessary to form a group of folk dances that would revive the tradition. So, in 1967, a group of young people got together for the first time to rehearse a choreography of three folk dances. In subsequent years, the group performed at all the Festival openings (PREFEITURA MUNICIPAL, 2017).

In January 1978, the Feliz German Folk Dance Group was officially founded. Thus, activities began to develop throughout the year, allowing performances in other cities.

Dr Théo Tássio Shlatter composed a song to liven up the Festival, which is enjoyed during the days when the Chopp Festival is held (ASSMANN, 2009).

Festival Anthem

In the sweet Vale do Caí
Smiling at us from the hills
Jewel of the colonial area
Beautiful, peaceful, ideal
Like the city of Feliz
Whose name says it all
And whose people are happy and good
Likes a German draught
When we want to forget
The sorrows of life
Let's try it at the Festival
Draft beer, legal beer!
It's good to dance
We declare our love
But the best thing is
A little beer for us.

It should be noted that during the period before the Chopp Festival was set up, the municipality began rehearsals with various choirs, the most prominent of which formed the Coral da Sociedade Cultural e Esportiva Feliz, which began on 24 August 1973, under the direction of conductor Agostinho Normélio Ruschel (ASSMANN, 2009).

Another festival that recalls the traditions of this ethnic group is the Meeting of Craft Breweries and German Culture. The event aims to revive Feliz's Germanic and brewing tradition, which began in 1893 when João Ruschel founded the first high-fermentation brewery in Brazil and which, in 1934, was renamed Cervejaria Ruschel (MUNICIPAL PREFEITURA, 2017). (FIGURE 6).

The event's attractions include an exhibition of craft breweries, a band festival, a senior citizens' ball, artistic performances and traditional bands. There will also be colonial delicacies and typical German cuisine, as well as an orchid exhibition and a craft trade.

Still on the subject of festivities, the National Blackberry, Strawberry and Whipped Cream Festival - FENAMOR - stands out in the municipality. Every even-numbered year, the Feliz Municipal Park hosts this festival, focusing on tourism and publicising the municipality's agricultural, industrial and commercial potential, as well as Germanic culture and traditions (FIGURE 6).

This great event takes place in the middle of the strawberry and blackberry harvest. The festival was held in 1991, 1993, 2001, 2002, 2003, 2004, 2005, 2006, 2007 and 2008. By decision of the Feliz community, FENAMOR will now be held every two years (MUNICIPAL HALL, 2017).

The main objective of the festival is to revive and promote colonial traditions, which includes typical German music and cuisine, a fair of colonial products and handicrafts. The programme features national attractions, a ball for senior citizens, artistic and cultural performances, an industrial and commercial exhibition, family farming, a display and selection of blackberries and strawberries picked in the town, gastronomy workshops, among other activities.

In addition to the typical festivities that characterise the Germanic people in the municipality, it's worth highlighting other meetings that are promoted. These events have an intimate connection with the reality of the municipality, which is based on aspects that are reminiscent of the Germans.

This event is promoted by the local population and is open to all categories of antique and related vehicles (cars, motorbikes, utility vehicles, lorries and buses) over 25 years old, both domestic and imported. The event also contributes donations to the Schllater Hospital[15] in Feliz, making it a great attraction for tourists from all over the state. (MUNICIPAL PREFEITURE, 2017). (FIGURE 6).

And finally, the other event is the Cycling Tour, which promotes interaction with the countryside paths, the beautiful streets of the urban area and even trails in the middle of the native forest, which attract hundreds of *Moutain Bike* enthusiasts. The events organised by the community for this public are among the largest in the state in this category and are already listed as the main events in the territorial unit under study (FIGURE 6).

---

[15] In 1909, the Austrian Gabriel Schlatter (1865-1947) bought and remodelled a building erected in the 1890s to set up a hospital there, with a capacity for 25 beds. The history of the hospital is the story of three generations. The existence of an excellent hospital helped make Feliz rank first among the Brazilian municipalities with the highest human development index (HDI) in 1998. Memory manifests itself through people, places, buildings and objects, all with their own values and meanings. This space is full of stories and reflections that the municipality of Feliz receives and includes in its tourist and cultural itinerary (PREFEITURA MUNICIPAL, 2017).

The municipality of Feliz stands out on the cultural scene for its typical German festivals. The traditions that originated in the mother country of Germany led these people to settle in the territory of Rio Grande do Sul, materialising their knowledge and traditions. Thus establishing the teachings passed down through the generations, which are reflected in their typical German festivals.

These Germanic cultural issues are very significant in the spatial organisation of Feliz, making it recognised for its numerous cultural manifestations. In this sense, the sense of belonging of these people is remarkable, as they take pride in keeping their cultural traditions alive, passing them on from generation to generation. This is evidenced by the strengthening of this cultural group through its symbologies, materialised through the material code of culture, the typical festivals.

Figure 6 - Mosaic of photographs depicting festivities in the municipality of Feliz/RS
Source: Prefeitura Municipal de Feliz/RS (2017).
GOMES, L. C. (2017).

Another code to highlight is religiosity. It should be noted that, although religiosity is present in the cultural formation of the other municipalities that were colonised by Germans, in Feliz it can be seen in the Catholic and Evangelical churches (FIGURE 7).

The predominant religion is Catholicism. The town[16] de Arroio Feliz[17] in 1955 has its first chapel called Três Mártires Rio-.

Grandense. The current parish priest is Father Emílio Lunkes. The cemetery is Catholic and is located next to the Society (FIGURE 7).

In the town of Bananal[18] , which used to be called Picada Cará, the majority of the community is made up of evangelicals. For this reason, there is an evangelical church that was built in 1927 (ASSMANN, 2009).

Of particular note is the town of Bom Fim[19] , whose history is marked by the struggles between the first immigrants and the indigenous peoples who lived in the region at the time of colonisation. As for religion, there are more Catholics than Evangelicals.

Another town is Coqueiral[20] , which began around 1846 with the arrival of the first immigrants. These immigrants were Catholics. The only church in the town is the Evangelical Church of the Lutheran Confession of Brazil, built in 1918. The first Catholic church built near Coqueiral was in Picada Cará, called São Miguel (ASSMANN, 2009). (FIGURE 7).

In this sense, the locality of Escadinhas[21] , the majority of residents are Catholic. St Joseph's chapel is the community church. In front of the church, there

---

[16] Any place, considered in terms of what it may have in particular, settlement. In the case of Feliz, the localities of the municipality are diverse and are characterised as ARROIO FELIZ, BANANAL, BOM FIM, COQUEIRAL, ESCADINHAS, SOBRA DA BELA VISTA, LINHA TEMERÁRIA, MORRO DAS BATATAS, NOVA CAXIAS, PICADA CARÁ, RONCADOR, SÃO ROQUE and VALE DO LOBO (MUNICIPAL HALL, 2017).

[17] The first name given to the town was Kaudenbach. Kauden was the surname of a local resident. As it was forbidden to speak German during the Second World War, this name was changed and the town was baptised Arroio Feliz, due to the stream that runs through the community (ASSMANN, 2009).

[18] The first name came from one of the first residents, Guilherme Kayser, who owned a valley full of banana trees. That's why today the town is no longer part of Picada Cará and has been renamed Bananal (ASSMANN, 2009).

[19] The story began with an indigenous man called Luís "Buga", who was captured by immigrants from the region. Known at the time as "Buga's Land", there were conflicts over land ownership, with Buga being the indigenous leader against the immigrants. Some time later, after lengthy talks, they reached an agreement, and the town was renamed Bom Fim. (ASSMANN, 2009).

[20] This small town was first called "Palmenthal" and was later renamed Coqueiral. Its name came from the fact that there were many coconut trees where the first immigrants settled (ASSMANN, 2009).

[21] The name Escadinhas comes from a geographical accident located on the left bank of the river Caí. It is a natural rock in the shape of stairs that stretches along the river, approximately 120 metres long and 30 metres high (ASSMANN, 2009).

is a meeting room that is used as a mortuary chapel[22] and as a classroom for catechesis and other meetings. At the back of the chapel is the local cemetery.

In 1928, the town of Picada Cará[23] received the construction of St Michael's Church. It is reported that this was due to the large number of Catholics in the area. It should be noted that a large number of people from this locality followed religious life, becoming priests or nuns (ASSMANN, 2009). (FIGURE 7).

It is worth highlighting the town of Roncador[24] , which originated around 1870 with the arrival of the Germans. Religious life has always been present in the routine of its residents. On 25 May 1952, the São José Catholic Church was inaugurated. Most of the residents are Catholics, while the Evangelicals hold their services in the local school (ASSMANN, 2009).

In São Roque[25] the religiosity of the residents is also present. The predominant religion is Catholicism, with few adherents to other religions. The community has its own church, the construction of which began in 1927 and was completed in 1947. It has become a beautiful Germanic-style church (ASSMANN, 2009). (FIGURE 7).

Vale do Lobo[26] is also a locality where Catholics and Evangelicals predominate, with a small number of Catholics. The evangelicals began building their church on 8 November 1987 and completed it in 1996.

It should be noted that Feliz has both the Catholic Church and the Evangelical Church of Feliz. The history of these churches stretches from 1875 to 1907, with the construction and inauguration of their parishes. It should be noted that for many years there were only two religions, but during the period of industrialisation, there was a large influx of migrants, who brought other faiths

---

[22] Room where people's wakes take place.

[23] The first name given to the town was "Thawaks Tahl", because there were large tobacco plantations at the beginning of German colonisation. The name was later changed to Picada Cará.

[24] Roncador got its name from a waterfall that makes a lot of noise in times of flood, imitating the sound of snoring. In addition to this, many monkeys lived here and made these sounds (ASSMANN, 2009).

[25] The first name of the town of São Roque was "Bohnenthal", because it was said that one of the first immigrant families had the surname "Bohn". The name São Roque was in honour of one of the three martyrs of Rio Grande do Sul: Roque Gonçalves. (ASSMANN, 2009).

[26] The town's first resident was a German immigrant whose surname was Wolf, which means Wolf. As it is situated between two hills, the name Vale do Lobo came about (ASSMANN, 2009).

to the municipality. (CITY HALL, 2017). (FIGURE 7).

The first Catholic church in the region was built in the town of Morro das Batatas, which now belongs to the municipality of Feliz. The Evangelical religion began to spread in 1858, when the faithful were visited by a pastor from São Leopoldo or Campo Bom to perform religious services in private homes (ASSMANN, 2009).

It should be emphasised that the religiosity of the people of Feliz is currently distributed as 80% Catholic, 16% Evangelical and 4% belonging to the other churches present in this territorial unit (MUNICIPAL HALL, 2017).

In this way, the cultural code of religion in the municipality studied becomes evident, and this code is materialised in the ethnic-cultural landscape established by this group, evidenced by the large number of churches and religious groups that have been perpetuated over time in their various locations in the municipality of Feliz (FIGURE 7).

Figure 7 - Mosaic of photographs of churches that materialise religiosity in the municipality of Feliz and localities
Source: Municipality of Feliz (2017).
GOMES (2017).

Continuing with the peculiarities of German culture in the municipality of Feliz, and, along with it, the perpetuation of cultural codes, orality should be highlighted in the territorial unit, standing out as an intangible code of German culture in the municipality studied.

In this way, we can emphasise that the first forms of transmission of human knowledge were oral. Calvet (2011, p. 140) states that all contemporary literate societies were, at some point in their history, societies with an oral tradition, the remnants of which can be found in everyday or children's linguistic practices, such as proverbs and swear words.

In this sense, we can emphasise that orality is present in the daily lives of these people, who keep the traditions of their ancestors alive. This can be seen in the schools of Feliz. The town is an example in terms of education throughout the country. German classes are held in classrooms, where the language is taught as a way of valuing and preserving German culture, with the aim of giving students an edge in the labour market and valuing their mother tongue.

The municipality's municipal education system offers students German lessons from the 1st to the 9th year of primary school. The teaching of German is very important for the region, as students who speak and understand German have more advantages when looking for a job in the municipality. It also becomes very important when the student is studying at a university and decides to go on an exchange programme.

It's worth noting that most of these students, as well as the majority of the population of Feliz, speak the Germanic dialect. This significantly facilitates learning the standard German offered and taught in the town's classrooms.

This contextualises the fact that orality is an important intangible code in the territorial unit under investigation. The prospect of preserving the mother tongue remains present and alive in the municipality's population, which is transmitted and perceived in shops, schools or by simply walking through the streets of the municipality of Feliz.

CHAPTER 5

FINAL CONSIDERATIONS

> If there is someone who, par excellence, could never and should never complete a work, it is the author (RAFFESTIN, 1993, p. 266).

The need to preserve cultural heritage is a relatively recent issue in Brazil. In this way, discussions about the inherent aspects that structured the immigration process in the national territory have established peculiar marks in space, especially in culture and the economy. The importance of valuing the past and collective memory shapes colonisation as a remarkable process of cultural exploration/implantation.

In this sense, by carrying out this work on the peculiarities of the cultural contributions of the Teuto-Brazilian immigrants, as well as examining their migratory genesis in the municipality of Feliz, the materialisation of this cultural group in space became evident.

It can be emphasised that the aforementioned cultural manifestations of this group are responsible for the cultural identity of its people, since this identity is inherited from their ancestors. This factor indicates that the cultural elements of German immigrants maintain the spatial uniqueness of the territorial unit analysed, establishing cultural codes that are preserved and materialised in the local landscape.

This work highlights the contribution and perpetuation of German culture, since the material codes are present in the history of these people. In other words, data has been obtained that is consistent with Feliz's migratory process, as well as its ethnic cultural formation.

It is worth noting that by carrying out the fieldwork, together with the historical recovery and the ethnic contribution, it was possible to observe the cultural marks of these Germanic immigrants, allowing us to analyse and understand the various cultural elements present in the landscape of the area analysed. The perpetuation of architectural, festive and religious traits refers to the specific material codes of this culture.

It should be noted, then, that typical German architecture is materialised in Feliz's landscape as an important cultural code. The façade of the town hall and the town's half-timbered buildings and houses stand out. The façade of the town hall stands out as a tourist and historical attraction, making it possible to see the typical architecture of the German-Brazilian people preserved in the territorial unit under study.

In this way, this code establishes itself as one of the most expressive in the municipality's landscape, where the Germanic architectural style, the Enxaimel style, is materialised on the various façades of the houses in Feliz. Thus, the material cultural code becomes essential for the ethnic identification of this group.

Another material code that is evident in Feliz is the typical German festivals, which are characterised by the festivals that are held in the municipality throughout the year. They allow German colonial traditions to be revived, including typical music and gastronomy. The fair of colonial products and handicrafts is also a highlight.

Still on the subject of festivities in Feliz, FENAMOR stands out. It promotes a display of typical German cuisine, emphasising dishes that are typical of this culture. Among these dishes we can highlight the presence of cucas, biscuits, sausages, jellies, among others. This cuisine characterises the gastronomic meeting that takes place annually in the municipality as a mark of the local German-Brazilian ethnic group.

It should also be noted that religion is present in the territorial unit under study, materialised in the various churches. This registers the material code of ethnic Germans in the municipality. The perpetuation of this code brings alive the religious importance of this social group, where Catholics and Evangelicals make up a significant number of churches that have perpetuated this code in Feliz and its rural localities.

With regard to the immaterial codes of Germanic culture in this territorial unit, orality stands out. It is alive in the daily lives of its people. This code is so important that it is still transmitted today in basic education in the public schools of Feliz. In addition, the use of the German language is still common among the inhabitants, on the streets, in the shops and in various places in the town. As such, the traditions of their ancestors are still alive.

The identification of the municipality's diverse cultural potential through its cultural codes is expressive in orality, architecture, festivities, gastronomy and religiosity. They therefore bear witness to the Germanic cultural traits present in the municipality.

This work made it possible to recover the genesis of the German migration process in the municipality of Feliz, as well as to identify the material and immaterial cultural codes of this ethnic group, verifying the contributions of German culture to the socio-spatial organisation of the territorial unit analysed.

At the end of this investigation, it was possible to realise the contribution of Germanic cultural codes, as well as to analyse and understand German cultural insertion in the local/regional context. This understanding considered the cultural perspective, its contribution and perpetuation to the historical/ethnic/cultural organisation of Feliz up to the present day.

CHAPTER 6

REFERENCES

ALMEIDA, M. G., Landscape diversity and territorial and cultural identities in the Brazilian hinterland. In: ALMEIDA, M. G CHAVEIRO, E. F., BRAGA, H. C. (Org.) **Geografia e cultura os lugares da vida e a vida dos lugares.** Goiânia: Vieira, 2008.p. 47 - 74.

ASSMANN, B. E. S. **Happy yesterday and today.** Porto Alegre: Corag - Companhia Rio-grandense de Artes Gráficas, 2009.

BEZZI, M. L. **Região:** Uma (re)visão historiográfica da gênese aos novos paradigmas. 1997. 337 f. Thesis (Doctorate in Geography) - Universidade Estadual Paulista, UNESP, Rio Claro, SP, 1996.

. **BRUM NETO. "Cultural Regions: the construction of cultural identities in Rio Grande do Sul and their manifestation in the gaucho landscape".** Sociedade e Natureza, Uberlândia, n.02, 2008. p. 135-155.

BRUM NETO, H. **The process of ethnic-cultural occupation and its influence on the organisation of the geographical space of the Geographical Microregion of Restinga Seca - RS.** 2004. 93 p. Coursework (Geography) - Federal University of Santa Maria, Santa Maria, RS, 2004.

______ . **Cultural Region:** the conception of cultural identities in Rio Grande do Sul and their manifestation in the gaucho landscape. 2006. 51 f. Master's thesis qualification (Master's in Geography) - Federal University of Santa Maria, Santa Maria, RS, 2006.

______ . **Cultural regions:** the construction of cultural identities in Rio Grande do Sul and their manifestation in the gaucho landscape. 2007. 319 f. Dissertation (Master's in Geography) - Federal University of Santa Maria, Santa Maria, RS, 2007.

______ . **The territories of German and Italian immigration in Rio Grande do Sul. 2012. 318 f. Thesis (Doctorate in Geography) - Universidade Estadual "Júlio de Mesquita Filho", Presidente Prudente, SP, 2012.**

CAETANO, J. N. **A influência cultural portuguesa na reorganização do espaço da microrregião geográfica de Cruz Alta/RS.** 2012. 270 p. Dissertation (Master's in Geography) - Federal University of Santa Maria, Santa Maria, 2012.

CALLAI, H. C. **O estudo do lugar como possibilidade de construção da identidade e do pertencimento**; VIII Congresso Luso-afro-brasileiro de ciências sociais - a questão do novo milênio, 2004.

CALVET, L. **Oral tradition & written tradition.** São Paulo: Parábola Editorial, 2011.

CLAVAL, P. **Cultural Geography.** Translation: Luiz Fugazzola Pimenta; Margareth Afeche Pimenta. Florianópolis: Ed. da UFSC, 1999.

______ . The role of the new cultural geography in understanding human action. In: CORRÊA, Roberto, L.; ROSENDAHL, Zeny. (org) **Matrizes da Geografia Cultural**. Rio de Janeiro: EdUERJ, 2001.

COLLEGE OF ARCHITECTS. **Architectural Terminologies.** 2017. Available at: <http://www.colegiodearquitetos.com.br/dicionario/>. Accessed on: 17 October 2017.

CORRÊA, R. L.; ROSENDAHL, Z. Cultural Geography: introduction to the theme, texts and an agenda. In: CORRÊA, R. L.; ROSENDAHL, Z. (Org.). **Paisagem, Tempo e Cultura**. Rio de Janeiro; Ed. da UERJ, 1999, p. 51-53.

______ . ROSENDAHL, Z. Cultural Geography: Introducing the Theme, the Texts and an Agenda. In: CORRÊA, R. L.; ROSENDAHL, Z. (Orgs.). **Introduction to Cultural Geography.** Rio de Janeiro: Bertrand Brasil, 2003. p. 918.

COSGROVE, D. E. Towards a radical cultural geography: problems of theory. **Space and Culture**, n. 5, p. 01-03, 1996.

______ . Geography is everywhere: culture and symbolism in human landscapes. In: CORRÊA, R. L.; ROSENDAHL, Z. (Org.). **Paisagem, Tempo e Cultura** ______ . Rio de Janeiro: Ed. da UERJ, 1998, p. 92-123.

______ . Worlds of Meaning: Cultural Geography and Imagination. In: CORRÊA, Roberto Lobato; ROSENDAHL, Zeny (Org.). **Cultural Geography: A Century (2)**. Rio de Janeiro: Ed da UERJ, 2000. chap. 02, p. 33-59.

______ . Towards a Radical Cultural Geography: Problems of Theory. In: CORRÊA, Roberto Lobato; ROSENDAHL, Zeny (Org.). **Introdução à Geografia Cultural.** 2. Ed. Rio de Janeiro: Bertrand Brasil, 2007, p. 103-134.

CUCHE, D. **The notion of culture in the social sciences**. Translation: Viviane Ribeiro. 2. ed. Bauru: Ed. da USC, 2002.

FERREIRA, Aurélio B.H. **Novo Dicionário da Língua Portuguesa.** 2 ed. Rio de Janeiro: Nova Fronteira, 1986. p. 361.

GABBI, J. V. **The cultural contribution of German immigrants to the history of Panambi: the main festivities and forms of leisure.** 2014. 47 p. Undergraduate Programme (History) - Universidade Regional do Noroeste do Estado do Rio Grande do Sul, Ijuí, 2014.

HAESBAERT, R. Identidades territoriais In: ROSENDAHL, Z.; CORREA, R. L. (Orgs.) **Manifestações da cultura no espaço.** Rio de Janeiro: EdUERJ, 1999, p. 169-190.

. Territorial identities: between multiterritoriality and territorial reclusion (or: from cultural hybridity to the essentialisation of identities). In: BEZERRA, A. C. A.; et. al. **Identidades e territórios**: questões e olhares contemporâneos. Rio de Janeiro: Access, 2007.

HALL. S. **Cultural identities in postmodernity.** Translation: Tomaz Tadeu da Silva; Guaracira Lopes. Rio de Janeiro: DP&A, 1997.

BRAZILIAN INSTITUTE OF GEOGRAPHY AND STATISTICS. IBGE. Immigration of German settlers in Rio Grande do Sul. **Brazil 500 years.** Available at: <http://brasil500anos.ibge.gov.br/#>. Accessed on: 10 June 2017.

MIKESELL, Marvin. Afterword: New interests, unresolved problems and persistent tasks. In: CORRÊA, Roberto Lobato; ROSENDAHL, Zeny (org): **Cultural Geography:** A Century (2). Rio de Janeiro: EdUERJ, 29-90, 2000.

MCDOWELL, L. The Transformation of Cultural Geography. In: GREGORY, D. et al. (Org.) **Human Geography:** Society, Space and Social Science. Rio de Janeiro: Jorge Zahar Editor, 1996.

PELLANDA, E. **Colonização germânica no Rio Grande do Sul:** Porto Alegre: Globo, 1925.

PESAVENTO, J. S. The immigrant in Rio Grande do Sul politics. **RS: immigration and colonisation.** DACANAL, J.; GONZAGA, S. (Org.). Porto Alegre: Mercado Aberto. 1980.

PREFEITURA MUNICIPAL DE FELIZ. Available at <http://www.feliz.rs.gov.br/web/>. Accessed on: 15 Nov. 2017.

RAFFESTIN, C. **Por uma Geografia do Poder.** São Paulo: Ática, 1993.

REGO, T. C. **Vygotsky**: uma perspectiva histórica-cultural da educação. 2. ed.

Petrópolis: Vozes, 1995.

RUVIARO, R. E. **Tourism and memoriality:** aspects of immigration architecture in Silveira Martins/RS - Brazil. 2011. 103 p. Dissertation (Professional Master's Degree in Cultural Heritage) - Federal University of Santa Maria, Santa Maria, 2011.

VOIGT, E. **Paisagem e Diversidade** Cultural: As Identidades Culturais das Distintas Etnias em Santa Maria/RS. 2013. 198 f. Dissertation (Master's in Geography) - Federal University of Santa Maria, Santa Maria, RS, 2013.

WAGNER, Philip L.; MIKESELL, Marin W. The themes of Cultural Geography. In: CORRÊA, R.L.; ROSENDAHL, Z. (org). **Introduction to Cultural Geography.** Rio de Janeiro: Bertatand Brasil, 2003. p. 27-62.

WEIMER, G. **Arquitetura popular brasileira.** São Paulo: Martins Fontes, 2005.

ZANATTA, B. A. **The Cultural Approach in Geography. Temporais** (ação) (UEG), v.1, p. 249-262, 2008. Available at: Accessed on: 30 Apr. 2011.

Buy your books fast and straightforward online - at one of world's fastest growing online book stores! Environmentally sound due to Print-on-Demand technologies.

Buy your books online at
www.morebooks.shop

Kaufen Sie Ihre Bücher schnell und unkompliziert online – auf einer der am schnellsten wachsenden Buchhandelsplattformen weltweit! Dank Print-On-Demand umwelt- und ressourcenschonend produziert.

Bücher schneller online kaufen
www.morebooks.shop

Printed by Books on Demand GmbH, Norderstedt / Germany